AF730853

Connecting the Bible and Mathematics through Algebra (9 – 12)

Connecting the Bible and Mathematics through Algebra (9 – 12)

Javier L. Taylor

Black Jack Publishing
Atkins, Arkansas

Editorial, Sales and Customer Service office

Black Jack Publishing
2284 Worthen Road
Atkins, AR 72823
blackjackpress@gmail.com

Find us on Facebook!

All Scripture quotations and references are from the King James Version of the Bible.

ISBN 10: 0615479359
ISBN13: 978-0615479354

How does mathematics relate to the Bible?

While the Bible is not intended as a math textbook, it is rich its use of numbers, shapes, and three dimensional structures. We need to realize that people have been using sophisticated mathematics since the beginning. God created us with a natural curiosity and desire to count, design, and build. We have the opportunity to explore how mathematics might have been used thousands of years ago.

What is algebra?

Historically, algebra has been a cornerstone in the study and development of mathematics. Algebraic thinking can be found deep in the bedrock of mathematics and is therefore vital to the general education of all students from the early grades through high school. Algebra is often described as generalized arithmetic; this however, can diminish its role in the classroom. Arithmetic is useful for looking at relationships between numbers but it is algebra that is needed to describe the relationships between things that are changing and or between things that are changing and those that are invariant. Algebraic thinking emphasizes the relationship between quantities and the way these change with respect to one another. Algebraic thinking requires that the recognition of patterns and relationships which reveal the underlying structure of the quantities.

Why is the study of algebra important?

The study of algebra is essential to the development of high order mathematical thinking. The development of algebraic thinking begins with the development of number. Young students can use algebraic thinking to illustrate mathematical properties, to show relationships between quantities and to solve problems.

My purpose in putting these activities together

The purpose of this book is to use the Bible as a context for discussing the algebraic tools and models accessible by available to a high school student. I believe it is vitally important that we use the Bible as a context to discuss spiritual things, like holy living, like our service to God, etc. However, that said, it is also vital that our children see the Bible as relevant in other areas of life. I think this is what God was telling Moses to relate to Israel in Deuteronomy 6: 6 – 8. We are to discuss the Bible in many contexts including math and science. When you look at the data concerning church kids when they go off to college, it is alarming to say the least. We must battle for the future of our children by presenting them with the Word of God in a way that answers their questions and challenges them to examine the world around them carefully in light of the Holy Scripture. I hope this book gives the Christian school teacher or the Christian homeschooler activities which can be used to present the Bible as the Living Word of God.

Why integrate the Bible into the study of mathematics?

Biblical integration is probably one of the most difficult tasks for a secondary mathematics teacher. When I was in the classroom I felt like my time was restricted and that many of the topics I needed to cover were not directly covered in the Scripture. Topics like factoring quadratics, solving logarithmic equations, or determining the derivative of an equation. I would often use Scripture readings in class so that I kept the Word of God in the class, but I always felt like I didn't connect the Scripture to the mathematics. Disconnecting the subject matter from the Scripture forces the students to compartmentalize their faith, something like "the Bible is okay in Bible class but it has no relevance in math class." When I read the 2010 Barna Group Research results, I was heartbroken. The result of compartmentalizing our faith is that many Christians today are scripturally illiterate. We have to face the fact that when someone challenges the Scripture they are challenging our faith. We have to teach our children to have an answer for those who ask about or even challenge their faith (1 Peter 3: 15). A Christian's faith and the Word of God cannot be separated and the Christian remain whole. The Christian has to trust the Bible in every single area. Jesus, himself, said "the scripture cannot be broken" (John 10: 35).

Where does the study of algebra fit into a mathematics curriculum?

In its "Guiding Principles for Mathematics Curriculum and Assessment" the National Council of Teachers of Mathematics states the following about algebra and its connection to the mathematics curriculum:

> *"Algebraic concepts and skills should be a focus across the pre-K–12 curriculum. The development of algebraic concepts and skills does not occur within a single course or academic year. An understanding of algebra as a topic is a course of study. As a collection of mathematical understandings develops over time, students must encounter algebraic ideas across the pre-K–12 curriculum. At the elementary school level, teachers help students be proficient with numbers, identify relationships, and use a variety of representations to describe and generalize patterns and solve equations. Secondary school teachers help students move from verbal descriptions of relationships to proficiency in the language of functions and skill in generalizing numerical relationships expressed by symbolic representations."*

About the activities

- **Linear Growth Models**

This activity is based on 2 Peter 1: 5 – 8. I choose this scripture to show that the Christian is to constantly grow and add to their faith. When Students first study how things growth, the idea of a constant growth rate or slope is one that is recognizable and can create much valuable classroom discussion. This activity allows the student to see linear function in a variety of ways. The student should see them in tables/lists; scatter plots, graphs, and equations. The students should also be able to recognize them in verbal descriptions.

- **Modeling the Price of Land**

This activity is based on Leviticus 25: 1 – 24 and the idea that the price of the land that was, for the most part, leased until the year of Jubilee. This land then would be priced based on the amount of time until the next year of Jubilee. The students will have to make use of their knowledge of linear functions to build models to price the land based on various constraints. This activity allows students to begin modeling real-world situations such as the price of land in the Old Testament. The student will be challenged to build several models and account for several variables.

- **Non Linear Growth Models I**

This activity is based on Genesis 48: 16. The blessing of Jacob on the sons of Joseph speaks of a kind of growth that differs from linear growth. The students will consider how populations grow. This activity looks at growth models represented by quadratic and polynomial models. The student should be able to recognize these using difference tables, scatter plots and graphs. The students will build several different quadratic and polynomial models and identify the various components of each model.

- **Non Linear Growth Models II**

This activity is also based on Genesis 48: 16. The mathematical idea here is that of exponential growth and decay models. Students will have the opportunity to explore what an exponential function looks like in a difference table as well as a scatter plot and graph. The student will build several different exponential models and identify the various components of their model.

- **How Much Energy?**

This activity is based on Matthew 14: 14 – 20. The mathematical idea is that of using mass-energy equivalence to make calculations. This activity allows students to create a model to describe the energy necessary to feed the 5,000 (men) and to create a reasonable model for the entire population present at the feeding of the 5,000 men mentioned in Matthew chapter 14. Students will use Einstein's mass-energy equivalence to model the feeding of the 5,000 and make comparisons of the energy they calculated to other natural and man-made occurrences.

- **How Fast Can an Angel Fly?**

This activity is based on Daniel 10: 12, the activity gives students the opportunity to calculate possible velocities for the angel Gabriel as he carried the answer to Daniel's prayer for understanding. The teacher can discuss the idea of an angel having a mass and the consequences of faster-than-light travel.

- **The Limit of a Function**

This activity is based on Leviticus 11: 45 and can be used as an introduction to mathematical idea of the limit. Students will look at where a function is undefined and discuss why it is undefined. Students will also discuss the "closeness" in terms of mathematics. The activity relates this to our call to be holy.

- **Equations of the Last Days**

This activity is based on several passages in the book of the Revelation and was designed to help students think about the meaning of a variable and to model situations when part of the information is unknown.

Table of Contents

Mathematical Goal:
To guide students to better understand linear growth models.

Student Learning Expectations:

- Students will use an x-y table to determine if a growth model is linear.
- Students will calculate and determine the meaning of the slope of a linear growth model
- Students will calculate and determine the meaning of the y-intercept of a linear growth model

Topic question:
How do things grow?

Cooperative Groups: 3-4 students per group	**Materials Needed:** Paper Pencil Graphing Calculator
Questions to Consider: How will students represent (show) their thinking? How will students communicate (tell) their thinking? How will students reason (explain) their thinking?	**Vocabulary:** Mathematical Modeling Constraints Resolution of models

Lesson Timeline:

Time	Activity	Notes
2-3 min	Read 2 Peter 1: 5 – 8 and discuss why this could be considered a linear model	A linear model can be constructed by taking a number and adding the same number to it repeatedly such as: The number 2 and adding 6 to it repeatedly 2+ 6 + 6 + 6 + 6 + 6 + 6… The same principle applies here we are to be adding to our faith
10 min	The students should discuss what makes a pattern or situation "linear."	A function can be shown to be linear with a difference table, a graph or an explanation that repeatedly adding the same amount to an initial amount can produce a linear pattern or function.
15 min	Group the students in groups of 3 – 4	Putting the students in groups allows the students to express and defend their ideas, as well as the giving them the opportunity to share their thinking about the constraints of the problem.
20 min	**Explore part 1:** Have the Students complete student sheet # 1 and share their answers and various models. If there are not a variety of methods then the teacher can supply some of the methods found in the answers section of this book. **Explore part 2:** have the students complete student sheet # 2 and share their answers and various models. If there are not a variety of methods then the teacher can supply some of the methods in the answers sections of this book.	
15+ min	Summarize: Each group will share out their graphs. What does the graph tell us? How easy are they to read? Do they have titles? Do they have labels? Can true data statements be made from the graph?	

Next Day Extension:

In this lab you will need a ruler, and several different types of plastic or paper cups that are stackable.

Procedure:

Step 1: Measure the height of one cup and record it on the data collection sheet
Step 2: Stack a cup on top of the first cup and measure the height of two cups.
Step 3: Repeat this procedure for the remaining cups.
Step 4: Determine the dependent variable and the independent variable
Step 5: Plot the points on the graph and use the ruler to make a line of best fit.

Number of cups	***Heights (in centimeters)***

Questions:

1. Is the graph a linear plot?

2. What is the slope of the plot?

3. What does the slope represent in the problem?

4. What are the dependent and independent variables?

5. Write a function that describes the height of the cups?

Dependent Variable: ____________________

Independent Variable: ____________________

Explore Student Sheet # 1

1. Determine if the following relationship is linear. How do you know?

X	Y
1	3
2	6
3	9
4	12
5	15
6	18
7	21

2. James has purchased a \$5 card from a local drive through restaurant. Each time James buys a soft drink it costs him \$.79. How many soft drinks can James purchase with his \$5 card? Make a graph of this situation and find an algebraic equation to describe it. Now write the equation in the $y = mx + b$ form and identify the meaning of each of the parts of this equation

   Y means __________________
   m means ______________________
   x means ____________________
   b means _____________________

3. Determine if the following pattern is linear. Create a table and a graph and find an equation that fits the situation.

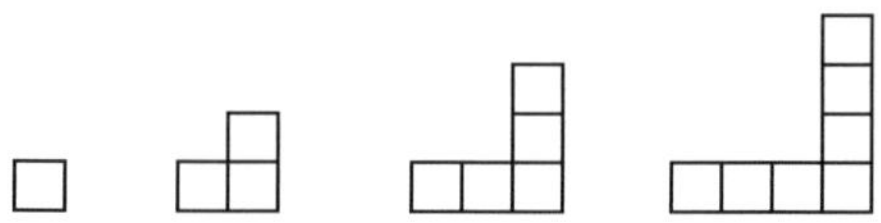

4. In the verse for this activity we are told to keep adding to our faith. Does this seem to indicate a linear relationship? Why or why not.

Explore Student Sheet # 2

1. Carole observed the following growth pattern in her Growing Crystals set over a four day period. Are the crystals growing in a linear pattern if so what is the pattern?

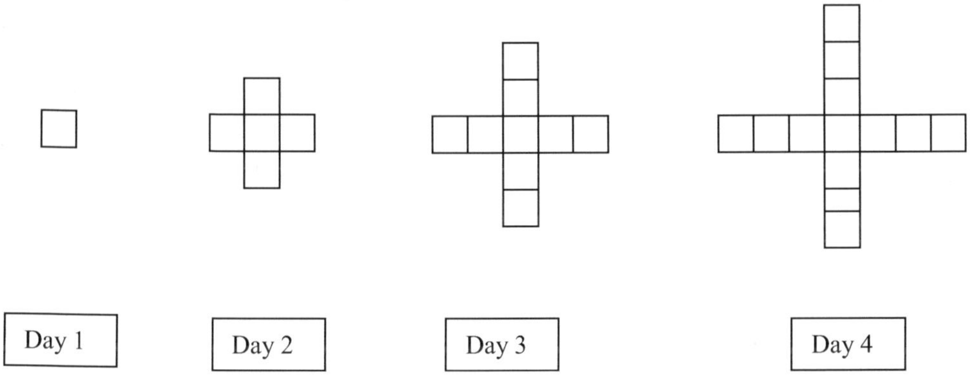

2. On day 4 the crystal is 7 units tall, how tall will it be on day 11?

3. On day 3 the crystal has 9 little pieces, how many pieces will the crystal have in 15 days?

4. Make a linear model to predict the number of units the crystal will have in t number of days.

Mathematical Goal:
To guide students to build an algebraic model to predict the price of land

Student Learning Expectations:
- Students will determine the variables of the situation
- Students will use a table or chart to list the variables
- Students will build a mathematical model to fit the situation

Topic question:
How much should the land cost?

Cooperative Groups: 3-4 students per group	**Materials Needed:** Centimeter grid Paper Chart paper Pencil Graphing Calculator
Questions to Consider: How will students represent (show) their thinking? How will students communicate (tell) their thinking? How will students reason (explain) their thinking?	**Vocabulary:** variable Model Constraint Cost Revenue Profit

Lesson Timeline:

Time	Activity	Notes
2-3 min	How much should that land cost?	Have the students read Leviticus 25: 1 – 24 and discuss how the system Moses was given could affect the price that land could be sold for.
5 min	Since all land reverted back to the original owners in the year of Jubilee, the closer you were to the year of Jubilee, the less you could get for the use of your land. If you purchased the land and build any structures on that land they would also belong to the original owner when the land reverted back to them. With this in mind we are going to build a chart that could be used to determine how much to sell land for depending on how long it was until the year of Jubilee.	The students will need chart paper and markers for this activity. First they need to determine how much land is worth per acre. For this they should use the internet and search land prices both city and rural. Most of the purchased was used either for livestock or for agriculture.
15 min	Group the students in groups of 3 – 4 and have them develop a table of land prices for up to 500 acres and a leasing price for up to 49 years.	The chart should be titled and labeled so that anyone could read it and make decisions based on the values found in the table.
30 min	**Explore part 1:** if the property is to be used for agriculture then find the price of seed to plant some combination of cotton, corn, wheat, barley, and hay. Determine the price per acre to lease the land and plant the crops. Have the students research the price of the crops	

	listed above per bushel or bale and determine the number of bushels or bales that can be reasonable expected per acre **Explore part 2:** when it comes to harvest time they need to determine how many people they would use per acre to harvest their crop the cost per hour is $10 for a laborer.	
15+ min	Summarize: Each group will share out their information and any models they might have used to determine cost per acre. What do the models tell us? How easy are they to use? Can they be used for any amount of acreage and any type of crops?	Bring out vocabulary when appropriate

Next Day Extension:

Figure cost and revenue equations for planting corn. You will have to make a plan for leasing the land to plant the corn. The plan should include the following: What kind of corn you plan to plant. How much you plan to spend on seed. How many acres you plan to lease and the number of years you plan to lease the property. Use the guides projected yield per acre. Also you will need to research the price per bushel so that you can sell you product at market price. Your teacher will tell you the price of leasing the property per year.

Once the research has been done write up your plan and present it to the teacher.

1. Write a generalized cost formula for a person looking to lease land for livestock. What are some possible expenses? What will be the source of revenue?

2. If you can sell your corn for $7.00 per bushel and your yield per acre is 130 bushels then build a model to show the income from your corn crops.

3. If your total cost per bushel were $4.68, and using the information in problem 2, build a profit function and determine the profit on 500 acres of corn.

4. The land you want to lease costs $2,000 per acre for the ten years and you lease 500 acres you get a 30% discount for leasing for 10 years. Figure your total cost for planting soybeans if the cost of the soybeans is $9.67 per bushel and you make $13.25 per bushel. Your yield per acre planted is 125 bushels. What would your profit be at the end of the 10 years? How can you insure that you will only have one Sabbath of rest for the land?

Mathematical Goal:
To guide students to better understand quadratic and polynomial growth models.

Student Learning Expectations:

- Students will recognize non linear growth from a difference table.
- Students will build non linear functions to model contextual situations.
- Students will identify the components of their models in the context of the problem.

Topic question:
How do things grow?

Cooperative Groups: 3-4 students per group	**Materials Needed:** Paper Pencil Graphing Calculator
Questions to Consider: How will students represent (show) their thinking? How will students communicate (tell) their thinking? How will students reason (explain) their thinking?	**Vocabulary:** Mathematical Modeling Constraints Resolution of models Regression

Lesson Timeline:

<table>
<tr><th>Time</th><th>Activity</th><th>Notes</th></tr>
<tr><td>2-3 min</td><td>Read Genesis 46: 16 and discuss what it means for "something" to grow other than in a linear context. What are some examples of this kind of growth?</td><td>A difference chart is a good way to determine if a relation is linear or non linear. Example:
<table>
<tr><th>X</th><th>Y</th><th>Difference in y values</th></tr>
<tr><td>1</td><td>4</td><td>2</td></tr>
<tr><td>2</td><td>6</td><td>2</td></tr>
<tr><td>3</td><td>8</td><td>2</td></tr>
<tr><td>4</td><td>10</td><td>2</td></tr>
</table>
Since all the first differences are constant then this is a linear relation. If the second differences were constant then the relation would be quadratic, and if the third difference was constant the relation would be cubic.</td></tr>
<tr><td>10 min</td><td>The students should discuss what makes a pattern or situation "linear." What can make a situation or pattern non linear?</td><td>The students really need to associate which difference is constant to the power of the relation.</td></tr>
<tr><td>15 min</td><td>Group the students in groups of 3 – 4</td><td>Putting the students in groups allows the students to express and defend their ideas, as well as the giving them the opportunity to share their thinking about the constraints of the problem.</td></tr>
</table>

20 min	**Explore part 1:** have the students complete the Explore 1 student sheet and discuss their answers. **Explore part 2:** Have the students complete the Explore 2 student sheet and discuss their answers. On both student sheets the students should produce a chart or graph as well as their models.
15+ min	Summarize: Each group will share out their graphs. What does the graph tell us? How easy are they to read? Do they have titles? Do they have labels? Can true data statements be made from the graph?

Next Day Extension:

Use the pictures below. There are four shapes in the pictures you can also refer to the table for the number of blocks in each shape.

Shape number	Number of blocks
1	1
2	5
3	14
4	30

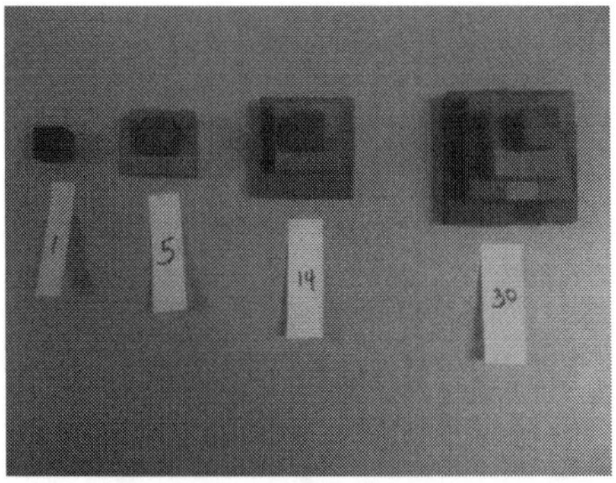

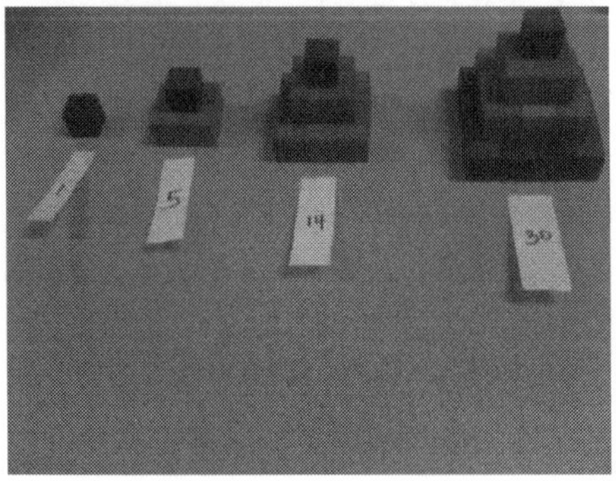

Find the number of blocks in the 10th shape of the series.

Can you find a rule for the number of blocks in any shape of the series?

Explore student sheet # 1

These are the first four shapes in a series of shapes

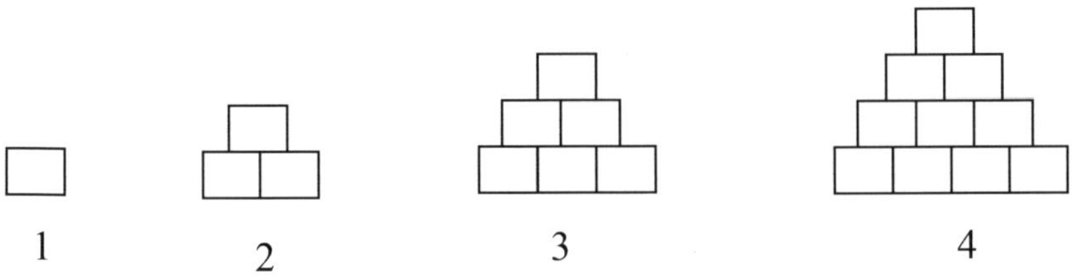

1. Find the number of blocks in the 5th shape.

2. Find the number of blocks in the 10th shape.

3. Find the number of blocks in the 25th shape.

4. Find the rule to determine the number of blocks for any shape in the sequence.

Explore student sheet # 2

These are the first four shapes in a series of shapes

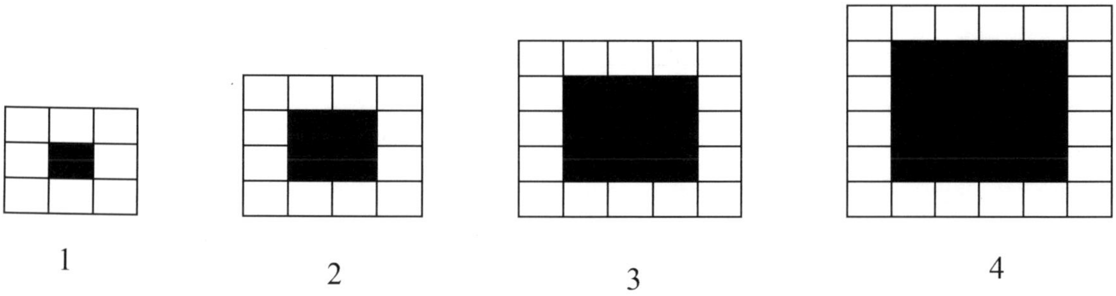

1. How many black blocks are in the 7th shape of the Sequence?

2. How many blocks in the border (white blocks) of the 10th shape of the sequence?

3. How many total blocks in the 15th shape of the sequence?

4. Find a rule for determining the number of blocks in the border of any shape in the sequence.

Mathematical Goal:
To guide students to better understand exponential growth and decay models.

Student Learning Expectations:

- Students will recognize exponential growth from a difference table.
- Students will build exponential functions to model contextual situations.
- Students will identify the components of their models in the context of the problem.

Topic question:
How do things grow?

Cooperative Groups: 3-4 students per group	**Materials Needed:** Paper Pencil Graphing Calculator
Questions to Consider: How will students represent (show) their thinking? How will students communicate (tell) their thinking? How will students reason (explain) their thinking?	**Vocabulary:** Mathematical Modeling Constraints Resolution of models Regression

<table>
<tr><th>Time</th><th>Activity</th><th>Notes</th></tr>
<tr><td>10 min</td><td>If you placed one penny on a corner square of a checkerboard and then doubled the number of pennies on each additional square, how many pennies would you so that every square had it "share" of pennies?</td><td>Have the students make a difference table for this problem.
<table><tr><th># of squares</th><th># of pennies</th></tr><tr><td>1</td><td>1</td></tr><tr><td>2</td><td>2</td></tr><tr><td>3</td><td>4</td></tr><tr><td>4</td><td>8</td></tr><tr><td>5</td><td>16</td></tr><tr><td>6</td><td>25</td></tr></table>When they keep getting a different difference, this is an indication that the function is exponential.</td></tr>
<tr><td>5 min</td><td>The students should discuss what makes a pattern or situation "exponential."</td><td>A function can be shown to be exponential with a difference table, a graph or an explanation that it grows by doubling, tripling, or by some power.</td></tr>
<tr><td>15 min</td><td>Group the students in groups of 3 – 4</td><td>Putting the students in groups allows the students to express and defend their ideas, as well as the giving them the opportunity to share their thinking about the constraints of the problem.</td></tr>
<tr><td>20 min</td><td colspan="2">Explore part 1: Have the Students complete student sheet # 1 and share their answers and various models.
Explore part 2: have the students complete student sheet # 2 and share their answers and various models.</td></tr>
<tr><td>15+ min</td><td colspan="2">Summarize: Each group will share out their graphs. What does the graph tell us? How easy are they to read? Do they have titles? Do they have labels? Can true data statements be made from the graph?</td></tr>
</table>

Explore Student Sheet # 1

1. Determine if the following function is linear or non linear.

f(x)	2	4	6	8	10
x	4	16	36	64	100

Determine the function and use it to find the value of f(15) and f(25).

2. This activity has been about exponential growth, list some things that grow exponentially.

3. A student opens an email which contains a virus. The virus accesses the student's email account and sends itself to the first 10 people on his contact list. Every time the email is opened it does the same thing to that person's computer. How many times will this process have to be completed to infect 1,000,000 computers?

4. In 1 Chronicles 21 we find that David sinned before the Lord by numbering the people. David chose the punishment of pestilence (disease) in Israel for three days, 70,000 people died of the pestilence. If God began this pestilence with two people who each infected two other people. How many times would this process have to happen to infect 70,000 people?

Explore Student Sheet # 2

I'm getting out of here!

You are trying the get out of the math classroom. The only rule is that you can only cover half the distance to the door at a time. Using the measuring tape, measure the distance to the door and record it in the table provided. Take half the distance for the second measurement and continue the process of measuring half the distance to the door until you can no longer measure. Record your data in the table.

Number of measurements taken	Distance from the door

1. Use your graphing calculator to create a scatter plot of the data in the table. Use L1 for the number of measurements and L2 for the distance from the door.

2. How is the distance from the door changing in relation to the number of measurements?

3. Considering the graph, will you ever reach the door? Why? Why not?

4. How many times did you measure the distance? What is limiting you from taking further measurements?

5. What is the regression curve?

Mathematical Goal:
To guide students to make conjectures, build mathematical models, and reason about their models.

Student Learning Expectations:

- Students will make and defend conjectures about the total number of people present at the feeding of the 5000 (men).
- Students will determine the mean serving size of fish and bread for each person present at the feeding of the 5000.
- Students will use the mathematical model $E = mc^2$ to determine the amount of energy created to feed the 5000 (men).

Topic question:
How much fish would it take to feed all those people?

Cooperative Groups: 3-4 students per group	**Materials Needed:** Paper Pencil Graphing Calculator
Questions to Consider: How will students represent (show) their thinking? How will students communicate (tell) their thinking? How will students reason (explain) their thinking?	**Vocabulary:** Mathematical Modeling Constraints Resolution of models

Lesson Timeline:

Time	Activity	Notes
2-3 min	Read Matthew 14: 14 – 20 and discuss the possible number of people present that day to hear Jesus preach	The Bible only records the number of men present who ate, however it is reasonable to assume that many of these men brought their families to hear Jesus preach. One way to start students thinking about a possible model is to ask how many people would have been there if half of the men present were married and brought their wives with them. Then add that half of those married had at least one child and brought their child with them.
10 min	The students should make some mathematical models of the possible number of people in the crowd who had followed Jesus. Discuss the constraints of the various models.	A function can be shown to be linear with a difference table, a graph or an explanation that repeatedly adding the same amount to an initial amount can produce a linear pattern or function.
15 min	Group the students in groups of 3 – 4 and have them determine the amount of fish in a normal serving. Also find the amount of bread found in a normal serving. If the information is found in ounces then convert the number to grams for use in the $E = mc^2$ model	Putting the students in groups allows the students to express and defend their ideas, as well as the giving them the opportunity to share their thinking about the constraints of the problem.

20 min	**Explore 1:** Using the information gathered by the groups, determine the mass (in grams) for the average meal that might have been consumed that day. (there are approximately 28 grams per ounce) The units will be Joules the students will also need to convert their mass measurements into kilograms. The speed of light in this calculation is $3 X 10^8$meters per second. **Explore 2:** Once the students made their calculations they need to have a point of reference for their findings. You can find the energy output in joules of an atomic bomb explosion, the output of the sun, etc.
15+ min	Summarize: Each group will share out their graphs. What does the graph tell us? How easy are they to read? Do they have titles? Do they have labels? Can true data statements be made from the graph?

Next Day Extension:

The students should develop a model to predict how many were present in the crowd at the feeding of the four thousand men mentioned in Matthew 15: 29 – 39. The students can then estimate the amount of energy created to feed these from the seven loafs and a few little fishes. The students can then make some mathematical comparisons for the amount of energy created by Christ.

1. Using the equation $e = mc^2$ calculate the amount of energy stored in your mass. (A pound is defined as .45359237 kilograms)

2. How much mass/energy would have to be created to feed 100 people a dinner if each person consumed 16 oz of food?

3. If $2.79X10^{20}$ Joules of mass/energy were used to create food and each person consumed 16 oz of food, how many people could be feed? Round to the nearest person.

4. Plot the mass/energy for 16 oz of food consumption for 10, 20, 50, 100, and 500 people. Find the regression curve of best fit. What does the slope of the regression line represent in this problem?

Mathematical Goal:
To guide students to make conjectures, build mathematical models, and reason about their models. Introduce students to the formulas for special relativity and have the students use these in context.

Student Learning Expectations:

- Students will estimate velocity based on extreme changes in distance and small changes in time.
- Students will use the formula for special relativity to solve problems
- Students will make comparisons between the calculated speeds and the speed of light.
- Students will estimate the amount of energy needed to accelerate a mass to near the speed of light.

Topic question:
How much fish would it take to feed all those people?

Cooperative Groups: 3-4 students per group	**Materials Needed:** Paper Pencil Graphing Calculator
Questions to Consider: How will students represent (show) their thinking? How will students communicate (tell) their thinking? How will students reason (explain) their thinking?	**Vocabulary:** Mathematical Modeling Constraints Resolution of models $KE = (1/2)mv^2$

Lesson Timeline:

Time	Activity	Notes
2-3 min	Read Daniel 9: 5 – 21 and discuss the amount of time Daniel spent praying before he was touched by the angel Gabriel. How fast do you think Gabriel would have to fly to reach Daniel before he had even finished his prayer?	The Bible records the prayer of Daniel in Chapter 9. Have a student read verses 5 – 19 aloud and time the reading this will give you a reference time for later calculations. It is uncertain how far Gabriel had to fly to meet with Daniel. If he indeed flew from the throne of God the distance he flew will be impossible to calculate we can only speculate.
10 min	The students should make some calculations based on Gabriel flying from the edge of our solar system, from the opposite edge of the Milky Way galaxy.	Have the students Google the distance across our solar system, the distance across the Milky Way, and the distance from M100. Why are these distances usually only estimates?
15 min	Group the students in groups of 3 – 4 and have them share their work is there agreement on the speeds they calculated? Have the groups share their work and discuss the possible ramifications of travel at faster-than-light speeds.	Putting the students in groups allows the students to express and defend their ideas, as well as the giving them the opportunity to share their thinking about the constraints of the problem.
20 min	**Explore:** Have the students Google the approximate distance from earth to galaxy M100 and calculate the speed Gabriel would have to fly to cover this distance in seven minutes time. Calculate the multiples of the speed of light this value represents.	
15+ min	Summarize: Each group will share out their graphs. What does the graph tell us? How easy are they to read? Do they have titles? Do they have labels? Can true data statements be made from the graph?	

Next Day Extension:

The question of whether angels have any mass plays a role in this problem. Assuming that angels have mass of 1 kilogram, calculate the energy necessary to propel Gabriel to 10% of the speed of light. Then use the same mass to find the energy necessary to propel Gabriel at the speed the class calculated for the distance across the solar system.

How much energy would be required if Gabriel had a mass of 100kg?

If an angel does have mass, would his mass remain constant as he reached speeds close to the speed of light?

1. If Gabriel were on the opposite side of the planet, how fast would he have to fly to get to Daniel in the approximately 7 minutes he was praying? (Hint: Google the diameter of the Earth and use ½ of this figure)

2. How close would Gabriel have to be in order to reach Daniel in seven minutes if he were traveling at the speed of light?

3. Google the speed of the fastest man-made objects. How much faster did Gabriel fly? (use the speed you calculated from the edge of the solar system)

4. Using the velocity calculated for Gabriel to travel from the edge of our solar system to Daniel (roughly 100 AU), calculate the time it would take a spaceship traveling that same speed to reach Proxima Centauri (approx 4 light years away).

Algebra

How Fast can an Angel fly? (Daniel 10: 12)

Mathematical Goal:
To guide students to informal definition of a limit

Student Learning Expectations:

- Students will determine when a function is undefined
- Students will create a table showing the function values close to a number
- Students will list some of the ambiguities of the term "close"
- Students will determine a limit value by simplifying a function

Topic question:
How close is close enough?

Cooperative Groups: 3-4 students per group	**Materials Needed:** Centimeter grid Paper Chart paper Pencil Graphing Calculator
Questions to Consider: How will students represent (show) their thinking? How will students communicate (tell) their thinking? How will students reason (explain) their thinking?	**Vocabulary:** Function Ration Undefined Limit of a function

Lesson Timeline:

Time	Activity	Notes
2-3 min	If you could approach the classroom door by covering half the distance from your desk to the door in one step and then continued this process could you ever walk through the door?	How close do you need to be to walk through the door?
15 min	The students measure the distance from a predetermined place in the room using meter sticks and then move half the distance from that spot to the door at a time. Have them make a table of this data (table provided) and make a scatter plot of the data.	The students are going to make a scatter plot of an exponential decay function. Since this function does not reach zero because they can always find half a measured distance no matter how small that distance is
20min	Consider the function $y = (x^3 - 2x^2)/(3x - 6)$ For what value of x is this function undefined? Why is the function undefined at this point? Construct a table of values choosing numbers greater than the number but close to the number. Create a table of values choosing numbers that are less than the number but close to the number. What do you notice about the tables?	The data for a standard rail car is in the excel file.
15+ min	Summarize: Each group will share out their graphs. What does the graph tell us? How easy are they to read? Do they have titles? Do they have labels? Can they make true data statements from the graph?	Bring out vocabulary when appropriate

Next Day Extension:

The Bible commands us to be holy as God is holy. Yet we live in bodies that seem to desire sin, and a world that seems to rejoice in sinfulness. What do you think God means then by the statement in Leviticus 11: 45 that we are to be holy? How can we use the idea of a limit to think about this?

Directions: Take a piece of 8.5" X 11" unlined paper and determine the area of the paper in square centimeters. Enter the calculated area in the "Area" column of the table below. Fold the paper in half and measure the area again and record the results. Continue to fold the paper in half until you can no longer fold the paper.

Folds	Area	Perimeter
0		
1		
2		
3		
4		
5		
6		
7		
8		

1. Graph the table above with the number of folds as the independent variable (x) and the area as the dependent variable (y). You can use a graphing calculator or the graph paper provided in this section.

2. Considering the graph, will the area ever reach zero? Why?

3. Will the perimeter ever reach zero? Why?

4. How many times did you fold the paper? Is there a limit to the number of folds?

Dependent Variable: ______

Independent Variable: ______

Mathematical Goal:
To help students define a variable and build an equation with more than one variable.

Student Learning Expectations:

- Students will define a variable in the context of problem
- Students will create an equation to calculate the population before, during, and after the Tribulation period.

Topic question:
How close is close enough?

Cooperative Groups: 3-4 students per group	**Materials Needed:** Centimeter grid Paper Chart paper Pencil Graphing Calculator
Questions to Consider: How will students represent (show) their thinking? How will students communicate (tell) their thinking? How will students reason (explain) their thinking?	**Vocabulary:** Function Variables Equation Expression

Lesson Timeline:

Time	Activity	Notes
2-3 min	How can you calculate a value if there are parts of the problem that are unknown?	The students need to have a deep understanding of the idea of variable in the context of a problem
15 min	Some of the examples of equations in context are: If there are 38 people in a room and 19 of them are female, then how many are male? Make an equation to demonstrate this situation. If the population of bacteria in a sample are doubling every hour, how can you know how many there are at a given time? Make an equation to model this situation.	The students should have little problem with the first example but may find the second situation a little more difficult. Have the students explain what part of the second situation is unknown. Then they can use the form $P = 2^x$ because time is the variable in this situation.
20min	What is going to be the population of people who enter the Millennial Reign of Christ? See if you can build a model of the situation using the following verses: 2 Thessalonians 2: 1 – 4; Revelation 6: 8; Revelation 8: 11; Revelation 9: 15; Revelation 11: 13.	The data for a standard rail car is in the excel file.
15+ min	Summarize: Each group will share out their graphs. What does the graph tell us? How easy are they to read? Do they have titles? Do they have labels? Can they make true data statements from the graph?	Bring out vocabulary when appropriate

Building the Equations of the Last Days

1. If the population of the Earth is **P** and the number of people who are raptured is **R** write an equation that models the population of the earth immediately after the rapture.

2. Revelation 6: 8 "**And I looked, and behold a pale horse: and his name that sat on him was Death, and Hell followed with him. And power was given unto them over the fourth part of the earth, to kill with sword, and with hunger, and with death, and with the beasts of the earth.**" How can the equation found for question 1 be modified to account for this?

3. Revelation 8: 11 "**And the name of the star is called Wormwood: and the third part of the waters became wormwood; and many men died of the waters, because they were made bitter.**" How can the equation found for question 2 be modified to account for this?

4. Revelation 9: 15 "**And the four angels were loosed, which were prepared for an hour, and a day, and a month, and a year, for to slay the third part of men.**" How can the equation found for question 3 be modified to account for this?

5. Revelation 11: 13 "**And the same hour was there a great earthquake, and the tenth part of the city fell, and in the earthquake were slain of men seven thousand: and the remnant were affrighted, and gave glory to the God of heaven.**" How can the equation found for question 4 be modified to account for this?

Answer Key for Extensions and Student worksheets

Modeling with Linear Functions next day extension:

1. If the same size cups were used and careful measurements were taken, then the function should be linear.

2. Answers will vary.

3. The slope represents the measure of the lip of the cup.

4. The number of cups is independent and the height of the cups is dependent

5. In general this will be h = measure of the lip of the cup (number of cups except the first cup) + height of the first cup minus the lip.

Modeling with Linear Functions Explore Student Sheet # 1:

1. This is a linear function because it has a constant rate of change.

2. James can buy 6 soft drinks and have $.26 left over. In the context of the problem **y** is the running total of what is left on the card, **m** is the cost of a soft drink, **x** is the number of soft drinks purchased, and **b** is the original amount of the gift card ($5 in this case).

3.

1	2	3	4	5	6	7	8
1	3	5	7	9	11	13	15

The equation is $y = 2x - 1$

4. It should work out as constant growth as we learn God's Word precept upon precept and line upon line, and we apply God's Word to

our lives. However, many Christians go through periods of little or no growth and other periods of great growth.

Modeling with Linear Functions Explore Student Sheet # 2:

1. The pattern is that after the first day the crystal grows by 4 units per day.

2. Since the height can be modeled by adding two each day except for the first day (2(n – 1) + 1), then the height on day 11 would be 2*10 + 1 or 21 units

3. Since the number of squares that make up the crystal can be modeled by adding 4 squares every day except the first day (4(n – 1) + 1), the number of squares on day 15 is 4*14 + 1 or 57.

4. The number of squares that the crystal contains is given as 4(n – 1) + 1 or can be modeled as 4n – 3.

Modeling the Price of Land next day extension:

Answers may vary depending on the amount of acres planted and the type of corn to be planted. The plans should include the number of years the land is to be leased, the type of corn to be planted, the cost of labor, and the expected profit. You set the lease price per acre. Use the following website as a guide:

http://www.extension.iastate.edu/publications/pm1885.pdf

Modeling the Price of Land Student Activity Sheet:

1. Profit (P) = income (I) – Expenses (E) or P = I – E

2. Income = $7.00 (130 (acres planted))

3. Profit is income minus expenses
   Since you have planted 500 acres and the normal yield is 130 bushels per acre then you have a total of 130 times 500 which is 65,000 bushels. Since expenses average $4.68 per acre then you have expenses of $304, 200 and income of $7 per bushel which totals $455,000 which yield a profit of $150,800. This can be achieved by 7.00 – 4.68 = 2.32 times 65,000

4. $2,000 per acre per year would total $1,000,000 minus the 30% discount would be $700,000. The total profit for ten years would be $1,313,750 (($13.25 – $9.67)(125 bushels per acre)(500 acres)(9 years)) – $700,000.

Non Linear Growth Models, Part 1 Next day Extension:

10th unit = $1^2+2^2+3^2+4^2+5^2+6^2+7^2+8^2+9^2+10^2 = 385$

Using the difference table you can see that the third difference is constant. This means that we are dealing with a cubic function, we can then use a graphing calculator to find the cubic regression.

Non Linear Growth Models, Explore student sheet # 1:

1. $(5 + 4 + 3 + 2 + 1) = 15$

2. $(10 + 9 + 8 + 7 + 6 + 5 + 4 + 3 + 2 + 1) = 55$

3. $(25 + 24 + 23 + \ldots + 3 + 2 + 1) = 325$

4. $n(n + 1)/2$

Non Linear Growth Models, Explore Student sheet: 2:

1. The 7th shape of this sequence would have 7^2 black squares or 49

2. The 15th shape will have $(15 + 2)^2 = 289$ total squares

3. The 10th shape of the sequence would have $(10 + 2)^2 - 10^2 = 44$ white squares

4. The border squares will have a total of $(n + 2)^2 - n^2$ white squares or $4n + 4$

Non Linear Growth Models, Part 2 Explore Student sheet # 1

1. No the function is not linear it is quadratic. The function is $y = x^2$
   $f(15) = 15^2 = 225$ and $f(25) = 25^2 = 625$

2. Example answers. Populations tend to grow exponentially, credit card debt, etc.

3. The entire process would have to be repeated 6 times.

4. The process would have to happen between 16 and 17 times with the actual answer being much closer to 16. ~16.09050673

Non Linear Growth Models, Part 2 Explore Student sheet # 2

1. Students complete the table in the calculator or other graphing utility

2. The distance to the door is ½ of the previous distance

3. Mathematically speaking, you can get infinitely close to the door without going through the door.

4. Students will usually reach a max of about ¼ to 1/16 of an inch, depending on the measurement tool they are using.

5. The regression curve is $y = (1/2)^x$

How much energy did God Create to feed the 5,000? Next day extension

It would be reasonable to estimate that at least ½ of the 4,000 men counted were married and it would also be reasonable to estimate that at least ½ of those married had at least two children so 4000 + 2000 + 2000 means the crowd could have been 8,000 or more.

How much energy did God Create to feed the 5,000? Student sheet

1. Since a pound is defined as having a mass of .45359237 kilograms then the mass of the student can be calculated by multiplying the weight in pounds by .45359237. My weight is 240 pounds so my

energy equivalence is $(108.862)(3x10^8)^2$ or about $2.19 \text{ X } 10^{19}$ Joules.

2. 100 times .45359237 is 45.359237 then the energy equivalence is about $4.08x10^{18}$ Joules

3. Since $2.79x10^{20}$ divided by $(3x10^8)^2$ is 3100 then the number of people who could be fed is 6,834.

4. 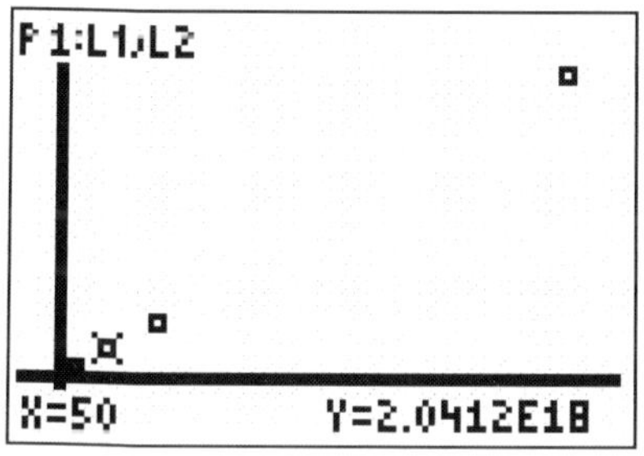

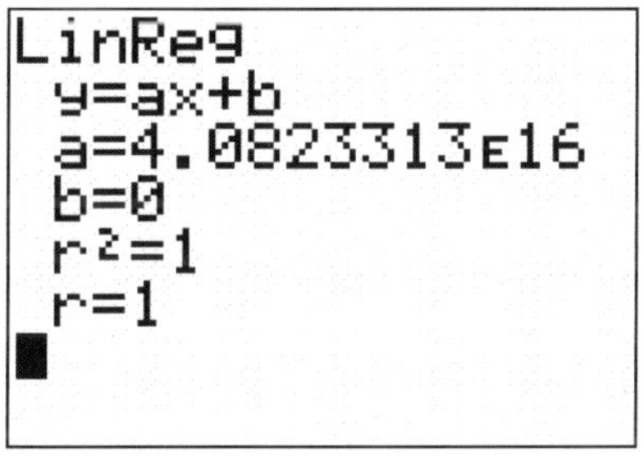

The slope of the regression line represents the amount of energy required to feed one person a 16 oz meal.

How Fast can an Angel Fly? Next day extension

The mass of a neutron is (1 kg) $((.1)3x10^8ms^{-1}))^2$ so the energy needed to accelerate him is $9x10^{14}$ Joules.

If Gabriel had a mass of 100 kg then the energy required to accelerate to 10% of the speed of light would be $(100kg)(.1(3x10^8ms^{-1})^2$ is $9x10^{16}$ Joules

If an angel does have mass then as speeds approached that of light there would be a relativistic mass that would have to be accounted for in the calculation.

How Fast can an Angel Fly Student Sheet

1. The earth has a circumference of about 40,075.017 km at the equator half of that is 20,0037.5085 km or 12,450.73 miles to travel this distance in 7 minutes you would have to travel about 106,720 miles per hour.

2. If Gabriel travelled at exactly the speed of light (186,000 miles per second) he could reach Daniel from a distance of 78, 120, 000 miles away in seven minutes.

3. The fastest man-made objects are the Helios probes launched in the mid-1970s Helios 2 had a perihelion velocity of 153,800 miles per hour. The distance from the Earth to the edge of the solar system is about 9 billion miles. To cover this distance in 7 minutes Gabriel would fly at about 7.714×10^{10} miles per hour. This is 501,579 times faster than the fastest man-made object.

4. The distance to Proxima Centauri is about 9.77×10^{10} miles and the calculated speed for Gabriel is about 7.714×10^{10} miles per hour. The time to get to Proxima Centauri would be 1.27 hours.

Limit of a Function next day extension

Student answers may vary

The idea of a limit is that you can get very close to a number without actually reaching that number. The same can be applied to our Christian walk, we may not actually be able to be holy (trapped in a sinful body) but we can more and more holy lives as we give over more control of ourselves to the Holy Spirit.

Limit of a Function student sheet

1. Students complete the table and plot it in a graphing utility

2. No the graph will get very, very close to zero, but since you can always take ½ of any distance it will never actually reach zero.

3. No, see answer above.

4. I was able to fold the paper 6 times and accurately measure the perimeter and area. The limit of the number of folds is related to the size of the paper, this was shown on Mythbusters, which can be looked up on Youtube and shown in class.

Other books in the Connecting the Bible and Mathematics Series

Connecting the Bible and Mathematics Through Data Analysis

Connecting the Bible and Mathematics Through Geometry

Connecting the Bible and Mathematics Through Number Theory

Connecting the Bible and Mathematics Through Measurement

Also Available from Black Jack Publications

Author and educator Javier Taylor Considers the topic of holiness as it relates the everyday life of a believer.

Made in United States
Troutdale, OR
11/03/2026

56849108R00046